TESLA MODEL X: DRIVING THE FUTURE

Luke Magic

Simone Magic

TESLA MODEL X: DRIVING THE FUTURE

Table of Contents

Chapter 1. Introduction

The Tesla Model X is not just a car - it's a vision of the future. With its advanced technology, innovative design, and eco-friendly features, the Model X is setting new standards for what a car can be. From its sleek exterior to its spacious interior, this vehicle is designed to offer an unparalleled driving experience that is both stylish and sustainable.

Since its founding in 2003, Tesla Motors has been on a mission to create electric cars that are not just environmentally friendly, but also practical, efficient, and exciting to drive.

With the introduction of the Model X in 2015, the company took a major step forward in achieving that goal. By incorporating cutting-edge technology and engineering, Tesla has created a car that is not just good for the environment, but also a joy to drive.

But the Model X is more than just a car - it's also a symbol of a larger movement towards a more sustainable and eco-friendly future.

With its advanced electric motor, regenerative braking system, and other innovative features, the Model X is leading the way towards a world in which cars are powered by clean and renewable energy sources.

In this book, we will take a closer look at the Tesla Model X and explore its many features, from its advanced autopilot system to its luxurious interior. We will also examine the history of Tesla Motors and the company's mission to create a more sustainable future for all of us.

Whether you're a fan of electric cars, a curious driver, or just interested in learning more about one of the most innovative cars on the road today, this book is the perfect guide to the Tesla Model X!

Chapter 2. A Revolution in Design

When the Tesla Model X was first unveiled in 2015, it was clear that this was no ordinary SUV. With its sleek lines, cutting-edge technology, and unique falcon-wing doors, the Model X stood out as a symbol of innovation and progress in the automotive industry.

But what really set the Model X apart was its design. Every aspect of the vehicle's shape, materials, and features was carefully crafted to maximize its efficiency, performance, and comfort.

One of the most striking elements of the Model X's design was its falcon-wing doors. These doors opened upward like wings, providing easy access to the second and third rows of seats, even in tight parking spaces. The doors were also equipped with sensors that could detect the height and width of surrounding objects, ensuring that they would never accidentally hit anything.

Beyond its doors, the Model X's exterior was designed with aerodynamics in mind. The sleek, tapered body reduced drag and improved the vehicle's efficiency, while the expansive glass roof provided ample natural light and improved visibility for the driver.

Inside, the Model X was just as impressive. The dashboard featured a massive touchscreen display that controlled everything from the air conditioning to the sound system to the car's Autopilot features. The seats were made from premium materials and offered ample legroom and headroom, while the vehicle's spacious cargo area provided plenty of room for luggage, groceries, and other items.

But what really set the Model X apart was how it all came together to create a truly unique and groundbreaking vehicle. From its falcon-wing doors to its state-of-the-art battery technology to its advanced safety features, the Model X represented a new era of sustainable and stylish transportation.

And while the Model X may have been designed for the future, it was also designed to appeal to drivers of all ages. Whether you were a tech-savvy millennial or a seasoned baby boomer, the Model X's design and features were intuitive and easy to use.

As Tesla CEO Elon Musk said at the Model X's launch event, "We wanted to create a car that would not just be a means of transportation, but also a work of art." With the

Model X, Tesla achieved that goal, creating a vehicle that was as practical as it was beautiful, and as innovative as it was timeless.

Chapter 3. The Power of Electric

The Tesla Model X is more than just a beautiful and innovative vehicle. It's also a powerful symbol of the shift toward electric transportation, and the potential for a cleaner, more sustainable future.

At the heart of the Model X's power is its electric drivetrain, which uses advanced battery technology to power its motors. The Model X comes with either a 75 kWh or 100 kWh battery, depending on the model, and can travel up to 371 miles on a single charge.

But the Model X's electric power isn't just about range. It's also about performance. With instant torque and smooth acceleration, the Model X can go from 0 to 60 miles per hour in just 2.7 seconds, making it one of the fastest SUVs on the market. But the benefits of electric power go beyond performance. Electric vehicles like the Model X are also cleaner and more efficient than traditional gas-powered cars. They emit no tailpipe emissions, reducing air pollution and improving overall air quality. And because they don't rely on fossil fuels, they also help to reduce our dependence on foreign oil and mitigate the impact of climate change.

But the benefits of electric power don't end there. The Model X's battery can also be used to power homes and other buildings during power outages or peak energy demand periods, reducing the need for traditional power sources and helping to stabilize the grid.

With the Model X, Tesla has not only created a vehicle that's beautiful and innovative, but also one that's helping to lead the way toward a more sustainable and resilient future. As Tesla CEO Elon Musk has said, "We want to accelerate the transition to sustainable energy. It's important for the future of humanity." And with the Model X, Tesla is doing just that.

Chapter 4. Safety and Autopilot

Safety has always been a top priority for Tesla, and the Model X is no exception.

From its advanced safety features to its high-strength materials, the Model X was designed with the safety of its passengers in mind.

One of the key safety features of the Model X is its reinforced safety cage. This is made from high-strength steel and aluminum, and is designed to provide maximum protection in the event of a collision.

The vehicle also includes front and side airbags, as well as an advanced rollover protection system. But safety in the Model X goes beyond just the vehicle's structure.

The Model X is also equipped with a suite of advanced safety features, including automatic emergency braking, lane departure warning, and blind spot monitoring.

These features work together to help prevent accidents and keep passengers safe on the road.

The Model X also includes Tesla's Autopilot system, which uses advanced sensors and cameras to help drivers stay safe and in control while driving.

Autopilot includes features like adaptive cruise control, which adjusts the vehicle's speed based on traffic conditions, and lane keeping, which helps keep the vehicle in its lane on the highway.

Autopilot also includes features like self-parking, which allows the vehicle to park itself in tight spaces, and Summon, which allows the driver to remotely move the vehicle in and out of tight parking spots using their smartphone.

But perhaps the most impressive feature of Autopilot is its ability to learn and improve over time. As more data is collected from Tesla vehicles on the road, Autopilot's algorithms become more advanced and its capabilities continue to expand.

With the Model X and Autopilot, Tesla is helping to push the boundaries of automotive safety and technology, and setting a new standard for what's possible in a vehicle.

Chapter 5. The Future of Transportation

The Tesla Model X is more than just a vehicle - it's a symbol of the future of transportation.

With its advanced technology, sleek design, and commitment to sustainability, the Model X is helping to pave the way for a new era of transportation.

One of the most exciting aspects of the Model X is its potential to disrupt the traditional automotive industry. With its electric power and innovative features, the Model X is challenging the status quo and forcing other automakers to rethink their approach to vehicle design and manufacturing.

But the Model X is also part of a larger movement toward more sustainable and efficient transportation.

As more and more consumers become aware of the impact of traditional transportation on the environment, there is a growing demand for alternatives like electric vehicles and public transportation. The Model X is helping to meet this demand by providing a viable and attractive option for consumers who want a vehicle that's both stylish and eco-friendly.

And with the rise of renewable energy sources like solar power, electric vehicles like the Model X are becoming an even more viable option for those looking to reduce their carbon footprint.

But the future of transportation is about more than just electric vehicles. It's about creating a more interconnected and efficient system that can move people and goods more quickly and sustainably. This includes innovations like autonomous vehicles, high-speed rail, and smart city infrastructure.

Chapter 6. The History of Tesla Motors

Tesla Motors was founded in 2003 by a group of engineers who wanted to prove that electric vehicles could be better than gasoline-powered cars.

The company's co-founder, Elon Musk, had already made a name for himself as an entrepreneur and visionary in the tech industry, having founded PayPal and SpaceX.

The early days of Tesla were focused on developing a prototype for an electric sports car, which was eventually unveiled as the Tesla Roadster in 2008.

The Roadster was a game-changer for the electric vehicle industry, as it demonstrated that electric cars could be just as fast and powerful as gasoline-powered cars.

Following the success of the Roadster, Tesla set its sights on developing a more practical and affordable electric car for the mass market. This led to the development of the Model S, which was unveiled in 2012.

The Model S was a breakthrough for Tesla, as it combined cutting-edge technology with luxury design and performance.

Since the introduction of the Model S, Tesla has continued to push the boundaries of electric vehicle technology with the Model X SUV, the Model 3 sedan, and the Model Y crossover.

In addition to its cars, Tesla has also developed advanced battery technology and solar panels for residential and commercial use.

But Tesla's mission has always been about more than just creating electric cars. The company's goal is to accelerate the transition to sustainable energy and help mitigate the impacts of climate change.

To this end, Tesla has also developed energy storage solutions and is working on developing fully autonomous driving technology.

Despite facing numerous challenges over the years, including production delays and financial struggles, Tesla has remained at the forefront of the electric vehicle industry.

Today, the company is one of the most valuable car companies in the world, with a market capitalization that exceeds that of some of the biggest names in the industry.

Chapter 7. The Engineering of the Model X

The Tesla Model X is a marvel of engineering, combining advanced technology and innovative design to create a unique driving experience.

From its aerodynamic shape to its cutting-edge battery technology, every aspect of the Model X has been carefully crafted to maximize performance, efficiency, and safety.

One of the most striking features of the Model X is its falcon-wing doors, which open upwards instead of outwards.

This unique design was developed to allow for easier access to the second and third rows of seats, while also improving aerodynamics and reducing wind noise. The doors are equipped with sensors that allow them to detect obstacles and adjust their opening angle as needed, making them ideal for tight parking spaces.

Another key feature of the Model X is its all-wheel drive system, which provides superior traction and stability in all driving conditions. The Model X is powered by two electric motors, one in the front and one in the rear, which work together to deliver exceptional acceleration and handling.

The motors are also highly efficient, helping to extend the range of the car's battery. Speaking of the battery, the Model X features Tesla's most advanced battery technology to date. The car is equipped with a large lithium-ion battery pack, which provides a range of up to 371 miles on a single charge. The battery is also designed to be highly durable and long-lasting, with a warranty of up to eight years or 150,000 miles.

In addition to its impressive powertrain, the Model X also features a range of advanced safety features. The car is equipped with a suite of sensors and cameras that provide 360-degree visibility and enable a range of driver assistance features, including Autopilot. The Model X also features a reinforced body structure, crumple zones, and advanced airbag systems to protect occupants in the event of a collision.

Overall, the engineering of the Model X is a testament to Tesla's commitment to innovation and sustainability. By combining advanced technology with elegant design, the Model X has set a new standard for what an electric car can be.

Chapter 8. The Future of Electric Vehicles

The Tesla Model X is more than just a luxury SUV - it represents the future of transportation.

As electric vehicles become increasingly popular, they are poised to revolutionize the way we think about cars and mobility.

One of the key advantages of electric vehicles is their efficiency. Unlike gasoline-powered cars, which waste a significant amount of energy as heat, electric cars convert nearly all of their energy into motion. This means that they can travel farther on the same amount of energy, making them a more sustainable and cost-effective option.

Electric vehicles also produce fewer emissions than traditional cars, which makes them an important tool in the fight against climate change. By reducing our reliance on fossil fuels, we can significantly reduce our carbon footprint and create a cleaner, healthier environment.

As technology continues to improve, electric vehicles are becoming more affordable and accessible. Battery costs are rapidly declining, and charging infrastructure is expanding.

Many governments are also offering incentives and subsidies to encourage the adoption of electric vehicles.

In addition to these practical benefits, electric vehicles also offer a more enjoyable driving experience. Electric motors provide instant torque and smooth acceleration, making them a joy to drive. They are also much quieter than traditional cars, reducing noise pollution and creating a more peaceful environment.

As the market for electric vehicles continues to grow, we can expect to see even more innovation and advancement. From longer range and faster charging times to new and exciting design features, electric cars are poised to become the norm rather than the exception.

The Tesla Model X represents a bold step forward in this new era of transportation. By combining cutting-edge technology with elegant design and exceptional performance, it has set a new standard for what electric cars can be.

As we look to the future, it's clear that electric vehicles will play an increasingly important role in shaping our world.

Chapter 9. Driving the Model X

The Tesla Model X is not just a stylish and eco-friendly vehicle; it's also a joy to drive.

From the smooth acceleration to the responsive handling, this car is designed to offer an unparalleled driving experience.

One of the most impressive features of the Model X is its electric motor. Unlike traditional gas-powered engines, electric motors deliver instant torque, meaning that the car can accelerate from 0 to 60 miles per hour in just seconds. This translates to a smooth and powerful driving experience, making the Model X a thrill to drive.

Another key feature of the Model X is its regenerative braking system. This means that when you take your foot off the accelerator pedal, the car's motor converts the vehicle's kinetic energy back into electricity, which is then stored in the battery. This helps to recharge the battery and improve the car's overall efficiency, while also providing a smoother and more controlled driving experience.

The Model X is also equipped with a host of advanced safety features. From the autopilot system to the collision avoidance technology, this car is designed to keep you and your passengers safe on the road.

The autopilot system uses cameras, sensors, and radar to assist with driving tasks such as lane changes, parking, and braking. Meanwhile, the collision avoidance system can detect and respond to potential collisions, helping to prevent accidents before they happen.

Of course, the Model X is not just about performance and safety - it's also designed to be comfortable and convenient. With features such as the panoramic windshield, falcon-wing doors, and spacious interior, this car offers a truly luxurious and enjoyable driving experience.

Whether you're commuting to work, running errands, or embarking on a road trip, the Tesla Model X offers a driving experience that is second to none.

With its cutting-edge technology, advanced safety features, and impressive performance, it's easy to see why this car is quickly becoming the vehicle of choice for drivers who value both style and sustainability.

Chapter 10. Conclusion

The Tesla Model X is a truly revolutionary vehicle that has set new standards for electric cars.

With its sleek design, advanced technology, and eco-friendly features, it has become a favorite among drivers who are passionate about both sustainability and style.

From the early days of Tesla Motors, the company has been on a mission to change the way we think about transportation.

With the introduction of the Model X, Tesla has taken another step forward in achieving that goal.

By creating a car that is both practical and exciting to drive, Tesla has shown that electric cars can be more than just a "green" choice - they can also be a luxury experience.

With its impressive range, advanced safety features, and cutting-edge technology, the Model X is a car that is designed to impress.

But beyond its many features and capabilities, the Model X is also a symbol of a larger movement towards a more sustainable and eco-friendly future.

By choosing to drive a Model X, you are not just making a personal choice - you are also making a statement about your values and your commitment to a better world.

As Tesla continues to innovate and push the boundaries of electric car technology, we can only expect to see more exciting and groundbreaking vehicles in the years to come.

But for now, the Model X stands as a testament to what is possible when we combine passion, creativity, and a dedication to making the world a better place.

The End

www.ingramcontent.com/pod-product-compliance
Lightning Source LLC
Chambersburg PA
CBHW070523220526
45467CB00002B/814